Your Child Has Diabetes

by

Diane Cairns

Copyright © 2015, Diane Cairns

ISBN: 978-1-329-06386-0

Table of Contents

Introduction

It doesn't matter how old you are, or how many children you have, being a parent is an exciting, but scary journey. Being responsible for another human life is a huge undertaking! Now throw in a life-threatening, chronic illness and the experience takes on a whole new dimension.

Twenty-five years ago, my son was diagnosed with diabetes. The cause of diabetes is unknown. In the past few decades, new types of insulin have been created, great advances have been made in how we measure blood sugar levels and there are now a variety of ways to get insulin into the body. But at the moment diabetes cannot be prevented nor cured – only managed. While scientists continue to improve the methods of monitoring the disease and researchers tirelessly search for a cure, people with diabetes are left with the ongoing task of managing this relentless condition.

This book is not a medical guide on how to manage diabetes, but rather a guide on how to raise a child who has diabetes. Some of the biological knowledge or suggested treatments referred to in this book may have changed between the time of me writing it and you reading it, but the skills it takes to successfully raise a child with diabetes will not have changed.

Through trial and error, research and experience, I have learned some things I think every parent of a child with diabetes should know. This book is a compilation of my top ten survival tips. I hope these tips make some of your challenges a little less daunting and help you to avoid some pitfalls.

Diane Cairns

1. Caregiver vs Victim

Since most of you reading this book have a child in your life who has been diagnosed with diabetes, you will all know what I mean when I say there is absolutely no comparison to life before diabetes and life after the diagnosis. The day diabetes came into your life (affectionately known as D-day) is the day your life got turned upside down. From that moment on, nothing has been simple.

When my son, Michael, was first diagnosed, we spent five days in the hospital. I was inundated with information. Not only did I need to learn a million things but I needed to be able to teach those things to everyone in Michael's world (teachers, babysitters, relatives, parents of his friends, etc.). Everything takes on a life of its own: school trips, play dates, birthday parties, Halloween, vacations – oh my! At one point there were ten adults in the room all frantically trying to teach, learn, understand, reassure, be intelligent, be competent, be brave... but then I just stopped. I looked over at the hospital bed in the corner and my world stood still. Michael was there, quiet and small, looking frightened and vulnerable. We were all busy trying to learn everything we needed to be able to care for this child. In our overwhelmed, panicked state, we forgot one very important thing. Michael. He was the one getting the injections, getting his blood tested, being made to eat or not being allowed to eat and not understanding why any of this was happening.

I knew that if we were going to survive a life sentence of diabetes intact, I needed to give my beautiful little boy some sense of control over his situation. I went over to him, looked him in the eyes and told him that we were all in this together. I explained that his pancreas was struggling and we all needed to learn as much as we could, so that we would have the ability to support an organ that was having trouble. This is the most important survival tip I can share with you. Give your child a positive role. Redefining the disease to be an 'ailment of the pancreas' rather than an 'illness of Michael' was a very powerful distinction. **By reframing the situation, Michael became a caregiver, not a victim.** Instead of being a 'sick kid', Michael was a healthy, intelligent, strong young man who was part of a team caring for his pancreas.

Just Sharing –

Being a caregiver flowed into other areas of Michael's life. At a very young age, Michael volunteered at a rehabilitation centre for orphaned wildlife. Having spent many years around needles and blood, he was comfortable around injuries and

*medical procedures. He would console and calm the birds
while the staff dealt with the injury.*

*Another example of his caring nature was when our 9-year-old
neighbour, Craig, lost his leg to cancer. The amputation took
place a few days before Halloween. While we were in the
hospital visiting, Michael and Craig started to talk about
Halloween. Michael explained that he wasn't going to do
Halloween because he couldn't eat any of the candy he
collected. But then he suggested "If you want, I can dress up
in your costume and collect the candy for you. I can do the
walking and you can do the eating. Between the two of us, we
can make Halloween work!"*

Whether Michael was a caregiver by nature or by
circumstance, I don't know. But I do know that being a
caregiver allowed Michael to play a powerful, positive role in
his own life.

2. Learn

As the parent of a child with diabetes, you will need to become an expert in the complex relationship between food, insulin and blood sugar. You will need to become proficient in medical procedures such as injections, blood testing and responding to life-threatening situations. You will need to become comfortable with numbers and mathematical equations, as you will be endlessly counting calories, weighing food, and adjusting insulin amounts.

I don't want to apply any unnecessary pressure, but you have to learn all of this in about five days. Oh, and by the way, it is usually in the same week of your life when you find out that your child has a chronic, life-threatening condition. The good news is that it can be done.

There are many books and websites to help you with all of this. There are support groups, on-line and in-person. Your doctor and your diabetic team will guide you to the right sources. Use all of them.

With diabetes, the name of the game is to keep the blood sugar level (BSL) within an acceptable range. In our world this range is between 4 and 8 mmol/l. Too low (less than 2.5) and you are dealing with seizures, comas or death. Too high (more than 12) and you are causing irreparable damage to various organs and systems in the body.

When I first started out, I took everything I read or was told as the absolute truth. I followed it to the letter. I measured food, portioned out a balanced meal, and timed the insulin peak to match the rate of digestion, only to find out that Michael's blood sugar was either too high or too low. How could that be? What did I do wrong? Did I not translate something correctly? Did he do something he wasn't supposed to do and not tell me?

If the equation was strictly 'this much **food** plus this much **insulin** equals this **blood sugar level**', you could get a book, do the math and live happily ever after. But alas, there is so much more to the equation. Diabetes does not exist in a sterile, static setting. It exists in human bodies and as such, the equation must factor in 'life'.

My tip to you is '**Don't rely on books, other people's experiences, or even logic. The best source of knowledge is your child**'. Watch and observe. His body will provide you with everything you need to know.

Your child will internalize different experiences from life and they will affect his BSL in a way that will be unique to him. This knowledge cannot be found in a book. A book may say 'running around the block uses up more energy than watching TV'. Biologically, we can all understand that. *But don't be fooled!* If the TV show Michael was watching was scary, exciting or a Stanley Cup playoff, he would burn up huge amounts of energy just sitting there and his BSL would take a beating!

Another thing to be aware of is that not all energy consumption is visible. Children are constantly growing. It is just the nature of the little darlings. A child may be quietly reading a book, looking happy and relaxed, but don't let your guard down – underneath that calm exterior a growth spurt may be getting ready to erupt. What is happening on the inside could be using up an incredible amount of energy.

When a young child is frightened or worried, the physical demands are as great as those of running a marathon. A teen going through a growth spurt, learning to drive a car and falling in love is burning up more energy than meets the eye.

Just Sharing –

One winter day, Michael woke up to find snow on the ground.
He got extremely excited and the adrenaline rush affected his
BSL. Walking to school in the snow took more energy than
normal because his little body had to work harder to keep
warm. Again his BSL was affected. A snowball fight at recess
used up a lot more energy than a rainy day recess activity.
Who knew snow was going to cause such a commotion in his
little body! Scientifically speaking, snow does not affect a BSL,
but in reality it really does!

You have to be very aware of your child's unique responses.
Every single second of every single day has the potential to
cause your child's BSL to spike up or down. It is impossible to
know or control all of the factors that can affect a BSL. But
the more you know about how different factors affect blood
sugar levels, the better you will be at living with diabetes.
Having Michael as the leader of his caregiver's team was very
helpful. He could give us insider information of what to watch
out for! This leads me to my next survival tip.

3. Know the Signs

Knowing the signs that indicate a high or low BSL and being able to respond to those signs is crucial to a long and healthy life.

Michael was diagnosed when he was six. He had two siblings; his older brother was eight, and his younger one was three. They were very close and spent a lot of time together. Michael's BSL would affect his mood and ability to cooperate and get along. If there was going to be any hope of a happy family life, we all needed to be able to respond to Michael's behaviour with awareness and intelligence.

In those early years, we tested Michael's blood four times a day: once before each meal and once before bed. When I say 'we', I really do mean 'we'. The four of us would gather around the kitchen table and try to guess what his BSL would be. If you guessed it correctly, you won a prize (hockey cards, erasers, Lego pieces, scratch and win lottery tickets) – nothing expensive, but something worth trying to get. This 'game' got everyone to take note and be aware of all of the internal and external elements that contributed to a BSL.

Diane Cairns

For example, Michael would say "I have a headache and I am thirsty so I think I am high. I guess 12.2". Another time, his older brother would look at him and say "You look pale, you have grey bags under your eyes and your hands are shaking. I think you are really low. My guess is 2.8". Even the youngest brother would pipe in and say "Michael was slow coming up the stairs and he got mad at me for no reason. I think he needs food. I am going to guess 3.5".

Testing Michael's BSL was a family affair. By making the testing a communal activity, we were once again reinforcing the idea that Michael was not alone. Together, we were all putting the pieces together and trying to understand how the human body reacts to life and the role insulin plays in that relationship. This rough-and-tumble group of little boys liked competing and winning; they took this game seriously. Everyone knew that Michael had an advantage because he actually knew how he was feeling on the inside. But since he was also the one who got his finger poked and had to supply the blood, no one ever complained about his advantage!

After playing 'Who can guess his BSL?' four times a day for months on end, we all got pretty good at knowing the signs. It taught all of us, including Michael, to use external signs and factors to gauge the internal state. This ability was invaluable over the years.

Just Sharing –

One very hot summer day, my husband, our three boys and I were in our car in the ferry line-up waiting to go to Vancouver Island. We were sitting in the middle of a sea of vehicles with no place to go. As you can imagine, boredom set in pretty quickly. The temperature was rising, the air was getting heavy, and the sides of the car seemed to be closing in. We had the windows open, but no air was moving. We all went from bored to grumpy very quickly. One punch, one slap, one complaint and then a whine. Michael started yelling; the little one started crying. I tried reasoning, negotiating, bribing, and threatening, and then started yelling myself. We then decided to close the windows as we were starting to attract attention. Then it finally happened. We could see in the distance that the line of cars had started moving. Yeah! We would be on the ferry in a few minutes and then these darling boys could get out and be in the fresh air. The boys must have sensed their torture was coming to an end because they had all calmed down. I turned to thank them when I noticed Michael was not calm. He was unconscious.

Diane Cairns

Oh my Gosh! I had been yelling at him to stop being a jerk, when in fact he was going low and heading towards a seizure. But I quickly realized that I couldn't take the time to berate myself at that moment. I needed to act. I quickly got into the back seat and started to get out the emergency kit. Then cars started honking. Our line was moving and we were holding up the show. So my husband put the car in drive and started to go. This sudden motion caused our youngest (who was sitting on the top of the back of the front seat) to fall. He started to cry.

This was only our second seizure. I wasn't really comfortable at handling them yet. I put honey in Michael's cheek and was waiting for something to happen. Nothing did! By now we were on the ferry. My husband wanted to get Michael to the First Aid station. On these ferries the cars are packed very close together. You can barely open the door to get an agile, thin person out of the car, never mind trying to pick up and carry out an unconscious child. But somehow we managed. Then we joined the hundreds of people gathering to walk up the very narrow and seemingly endless stairway to the upper deck of the ferry. I swear that if this mass of people were walking any slower they would have been going backwards. So there we were – me holding a crying 4-year-old, my husband carrying an unconscious 8-year-old, and my calm,

serious 10-year-old (bless him) all going nowhere in a
crowded, small, steep stairway in the middle of a ferry.

To make a long story short, we survived. The honey brought
Michael around, we fed him, and he was fine. By that
afternoon he was running around playing baseball with his
brothers, having a good old time.

Diabetes doesn't happen in a controlled science lab. It happens
out there in life with a million other things going on around
you. Everyone in that car knew the signs of a low BSL and we
all missed them! You may be educated, knowledgeable and
aware, but you are also human and you will make mistakes.
Just learn from each mistake and move on.

Diane Cairns

4. Educate Everyone

There is a saying, 'It takes a village to raise a child'. This is especially true when the child has diabetes. When their BSL is in the acceptable range, a person with diabetes is as normal as the next person. But... it only takes a moment to go from a normal BSL to a low BSL. When a person has a low blood sugar level, they may not have the physical or mental ability to correct the situation. When this happens, they must rely on others to assist them. You never know who your child may need to turn to for assistance. It may be a stranger on the street or a close friend.

The more people who know your child has diabetes, understand diabetes and know how to treat a low BSL, the better it will be for your child. It is your job to teach as many people as you can!

The first thing you need to do is let teachers, parents, friends, family, and neighbours know that your child has diabetes. This should not be done discreetly, nor 'when the time is right'. It should be done now! Everyone needs to know. Strangers (police, doctors, etc.) can learn that your child is diabetic through ID bracelets or necklaces, a sticker on a cell phone or a card in a wallet. But informing people close to him should be done in person. Once they know he has diabetes, they then need to know what diabetes is.

4a. What is Diabetes?

Diabetes is a complex condition. You, as the primary caregiver, will need to know about syringes, pens and pumps. You will need to understand the difference between fast-acting, short-acting and long-acting insulin. You will need to be aware of insulin peaks, digestion rates, ketones, hyperglycemia, hypoglycemia, glycemic index ratings, A1C levels, and....

When I first started out, I wanted to teach anyone who would listen to me. I wanted the whole world to understand what diabetes was. In my eagerness to do a good job, I went into too much detail and probably did more harm than good. People would get confused or bored and then just shut down. It didn't take me long to realize I had to tighten up my presentation. The people who may be called upon to save your child's life in an emergency do not need to know all the details. They just need to have a basic understanding of the disease, why things can go wrong and what to do to correct them.

I learned to describe diabetes in a language that was simple enough for children to understand and short enough for a group of parents to listen to while waiting outside a classroom door or at the side of a playing field. It goes like this:

> The human body digests food and turns it into energy (sugar/glucose) for all the organs in the body. Blood carries this energy to the various organs. Insulin is the key that unlocks the door that allows the energy to get out of the blood and in to the organs.

> In a normal person, insulin is produced by the pancreas and released into the body as needed. If a huge meal is consumed, the pancreas will secrete an appropriate amount of insulin. If nothing is eaten, no insulin will be secreted.

> But in a diabetic, the pancreas does not produce any insulin. Instead, insulin is injected into the body and is released over time. Therefore, food needs to be given to the body at a rate that corresponds with the insulin as it is being released. It is crucial to maintain an acceptable balance between the two. Too much or too little of either food or insulin will result in a blood sugar level that is either too high or too low.

The next part of your job as a teacher is to explain how a blood sugar level (BSL) affects a person's overall health and their immediate ability to function.

4b. Why are Blood Sugar Levels Important?

How a blood sugar level affects a person's ability to function is the most important thing you can teach. You must speak in a language that your audience can understand. I have found that most people can understand this explanation:

> All organs, including the brain, require a constant flow of good, healthy blood. When a diabetic is experiencing a low blood sugar level there is no energy (or sugar) in the blood. When the brain doesn't get enough energy it does not function well. The diabetic experiencing a low BSL may start to get confused and clumsy. If left untreated, they may start to get anxious and/or aggressive; from there the symptoms could move on to convulsions and even death.

(Note: I qualify my explanations in this book with 'may' or 'could', but when **you** are telling people, use the word 'will' and fill in the symptoms that your child actually displays.)

I have found that young children, or people who are not biologically-minded, grasp the concept of blood sugar levels better if I get them to visualize blood as being thick or thin. For example, if blood with a high BSL is visualized as blood that is too thick, it is easy to understand that this would cause damage to the small arteries found in the eyes, extremities (fingers and toes) and kidneys. A diabetic experiencing a high BSL does not feel well and future long term damage is being done, but death is not imminent. Conversely, if blood with a low BSL is visualized as being thin, it is easy to understand that thin, watery blood won't satisfy the needs of the brain. When the brain doesn't get enough substance, it freaks out, the body can't function. If left untreated, a very low BSL can be fatal within minutes.

Now that your audience has a basic understanding of diabetes and how high and low blood sugars affect the human body, the next step is to give them clear understandable instructions on how to respond to a low BSL.

4c. Treating a Low Blood Sugar Level

The reason I focus on how to treat a low BSL rather than a high BSL is because the time between a low BSL and death is not very long. A low BSL needs to be treated quickly by whoever happens to be close by. Conversely, complications and/or death from having a high BSL is a longer process. Very rarely would a stranger on the street or a friend on the playground be called upon to detect symptoms or provide treatment for a high BSL.

When you are teaching people how to treat a low BSL, you need to find the balance between making them appreciate the severity of the situation and not frightening them. Anyone near a diabetic may be called upon to provide life-saving assistance at any time. This task could seem rather frightening. You must make them feel competent and capable of handling the situation.

When teaching my caregivers, I like to explain that there are two levels of response. The first level of response is to recognize the signs and provide immediate assistance to prevent the BSL from getting any lower. The second level of response is to get the BSL back up to normal.

Children in general spend a lot of time surrounded by other children (classrooms, playgrounds, ice arenas, out in the backyard, etc.). Depending on their age, these children (classmates, friends and teammates) may be called upon to provide the first level of response. All they need to know is 'if a diabetic is acting strange or having trouble communicating, give them sugar, then get help'.

At this point there seems to be three questions that are commonly asked:

Q: What if I misread the signs and he is actually having a high BSL, not a low BSL?

A: It doesn't matter. As a first responder, always give sugar. If a diabetic is low, sugar will save his life. If a diabetic is high enough to be showing behavioural symptoms, a little higher won't matter.

Q: What do I do if there isn't any sugar around?

A: Most people with diabetes have sugar in various forms close by (i.e., sugar cubes in pockets, juice boxes in backpack, emergency food in tote bag...). So check these places first. Then, if necessary, knock on doors, stop a car – do whatever you need to do to get sugar!

Q: If a diabetic needs sugar can I give candy?

A: Candy is better than nothing, but it is not the best option. I don't use chocolate or candy to treat a low blood sugar level because it takes too long to get the sugar out of the fat in a chocolate bar and sucking on a hard candy or chewing a soft candy is just too slow. Sugar in any form is better than nothing, but if there is a choice, the most efficient form is liquid: juice or pop (but not diet pop!).

Once these first responders have interpreted the signs, and given sugar, their next step is to find help.

You need to make sure there are people nearby who are capable of providing that help.

When children are out in the world, there is usually a 'person in charge' nearby. This could be a teacher, a coach, a playground supervisor, or another parent. These people can be the second level responders. They will need to know how to get the BSL back to normal after the initial sugar has been ingested. And of course, it is up to you to teach them.

4d. Preventing a Relapse

I have found that people are better at following instructions if they understand the reasoning behind each action. I like to give the second level responders a quick lesson on how food is related to blood sugar levels.

This is how I explain the relationship between food and BSL to second level responders:

> Different foods are converted to sugar at different rates. Since a person who is experiencing a low BSL needs to get sugar into their system quickly, it is important that they ingest sugar that will be easily absorbed. Juice or pop (sugar in liquid form) gets absorbed very quickly, usually within minutes.

It is important to be aware of the fact that the faster a BSL rises, the faster it falls. For example, if a diabetic just has juice or sugary liquid, his BSL will rise quickly, but it won't stay there. It will fall down again just as quickly. So when treating a low BSL, it is important to give sugar first, but then it must be immediately followed up with food. To keep sugar coming into the system at a steady rate one should give a sugary liquid and then a starch and a protein. The sugar from the liquid will get into the blood stream within minutes, if not seconds. Just as that is used up, the sugar from the starch will become available, and shortly after that the sugar from the protein will become available.

In summary, an effective treatment for a low BSL is juice or pop (sugar) followed by cheese and crackers or peanut butter and bread (starch and protein).

4e. Create Competent Caregivers

When you live with diabetes, watching for the signs and knowing how to respond becomes second nature. But for those who don't know your child or only see him once a week or just at school, it is best to have things written down and have all the materials they will need to be successful close at hand.

I did this by having Diabetic Emergency Kits scattered throughout Michael's world. We had Emergency Kits in his classroom, his gym bag, the nurse's station, the teacher's desk, the staff kitchen, the hockey arena, the coach's bag, our neighbour's house, our car, his granddad's car, his uncle's boat... everywhere!!

The Emergency Kit was a small Tupperware container. On the outside was a printed information chart. This chart had emergency contact numbers, and symptoms of the different stages of a hypoglycemic reaction (low blood sugar level) and the corresponding treatments. The symptoms a person with diabetes experiences are unique to each individual. The list of symptoms on your Emergency Kit cover chart should be specific to your child, not the generic 'normal' ones listed in a book. Michael's first clue was that his legs would start to tingle. Then he would get quiet, confused and clumsy. Just before a seizure he would get agitated and aggressive, and then start convulsing. He could go from tingling legs to convulsing in about 15 minutes.

Each Emergency Kit had a juice box, an individual-sized snack pack of cheese and crackers or peanut butter and crackers, and a container of honey (the small pack you see at restaurants) or a package of glucose gel. In some kits (nurse's station and my car) I also had glucagon which could be injected.

The printed information sheet looked something like this:

Michael Smith (Insulin Dependent Diabetic)		
1234 Main St., Anywhere		Home: 604-123-4567
Mom cell:		Dad cell:
Dr. Jones:		Ambulance or 911
	Symptoms	Treatment
Stage 1	Tingling legs	Can treat himself with juice, cheese and crackers
Stage 2	Quiet, confused and clumsy	Can eat but will need help locating food and opening packages
Stage 3	Agitated and/or aggressive	Encourage him to drink and eat. Provide assistance if necessary.
Stage 4	Convulsing	*Put honey (or glucose jelly) inside cheek. Sugar from honey can be absorbed into the blood stream through the cheek membrane. Only do this if it is safe for you to do.
Stage 5	Unconscious	*Turn him onto his side and put honey inside his cheek. Call 911 or inject glucagon if you are competent at administering an injection.

*Note: On the GlucoGel package today it very clearly states, *"GlucoGel must never be given to unconscious individuals because of the risk of choking"*. Talk to your doctor about this.

20 years ago, I was told that if I found Michael unconscious, I was to put honey or sugar jelly in his cheek. It worked. But I have learned that rules change. By the time you read this book, things may have changed again. So check with your medical team and do what they recommend.

Each year I would replenish every Emergency Kit. This not only ensured the contents were fresh, but allowed me to make sure the kits were still where they were supposed to be. It also gave me an opportunity to speak with each teacher, coach, parent, etc. to remind them of the procedures and to answer any questions.

4f. Let Go

I taught people what they needed to know, in a language they could understand. I provided them with the knowledge and materials they needed to be successful in providing Michael with whatever form of assistance he might ever need. I couldn't do much more than that. I went to sleep at night knowing that I had prepared his world to the best of my ability.

Just Sharing –

Michael's class went on a field trip up the local mountain to go skiing for the day. I went as one of the parent chaperones. We knew that skiing and being cold would burn up a lot of energy. Our plan was that he would eat something between every run. We didn't take juice as it is messy when you fall on it. We didn't take crackers because they just turn to crumbs after a few falls. We had peanut butter and jam sandwiches stashed in every pocket available: my coat, his coat, his friends' coats.

We were prepared and had planned well. Right!

It was near the end of the day when the clouds came in quickly and the temperature plummeted. The wind picked up, making everything a bit more difficult. We decided this would be the last run. As we were skiing down, Michael stopped midway to eat his sandwich. Smart move. Good thinking.

While we were in line for the chairlift that would take us back up to the lodge, Michael admitted being tired and was glad to be finished. He ate the sandwich from my pocket. Michael and his friend went together in the two-man chair in front of me. As we went up the mountain, I was close enough to see them, but couldn't hear them. We were half way up when the lift suddenly came to a stop. Why? How long was it going to be before we started going again? After about 10 minutes I saw Michael eating the sandwich out of his friend's pocket.

I took a deep breath and did a little inventory of the situation. I could do nothing. I could see everything. Only a small bar was holding Michael into a chair that was suspended hundreds of feet above the ground. The wind was blowing and it was cold! He was tired and had no more food available to him. I had no idea how long the chairlift was going to be stopped.

I thought to myself `That is it – I quit! I can't do this anymore. I am never going to leave my house again`.

Wait – what was that? I heard something. I felt something ... the chairlift was moving again. Thank God!

A few minutes later we were all back in the lodge enjoying a big bowl of hot soup.

The moral of the story is 'Be prepared. Expect the unexpected. But in the end know that you can't control everything. Just remember to breathe'.

4g. Natural Safety Net

Before I conclude this chapter, I would like to take a moment to tell you something. In all my research on hypoglycemia (state of a person experiencing a low BSL) the description of the symptoms always ends with 'may cause' or 'could lead to' death. There are a few reasons why death is not a given. One is that if a low BSL is treated (i.e., the person is given some type of sugar), the BSL will start to rise immediately. A low BSL can be treated quickly and easily at every stage in its progression.

Another reason that death is not a given is because even if a diabetic gets to the point of being unconscious, with no one around to provide assistance, the body has one last safety net. The liver has a reservoir of stored glycogen. When blood sugar levels fall too low, the pancreas releases glucagon. Glucagon causes the liver to convert stored glycogen into glucose, which is then released into the bloodstream. The amount of glycogen stored in the liver is relatively small, but it may provide enough 'energy' for the diabetic to regain consciousness and locate food or get help. But take note that once this small reservoir is emptied, it takes food and time to replenish. So if you have a second hypoglycemic incident before the liver has replenished its supply, that safety net will not be there.

Just Sharing –

When Michael was a teenager, he had a seizure when he was alone. We know that he was unconscious for some period of

time and that he did 'come to' on his own and became aware
enough to call for help.

That night Michael's older brother quietly took a pillow and
blanket downstairs and slept on the floor beside Michael's bed.
He knew that Michael's liver wouldn't have had time to refill
its reservoir and that if Michael had a low BSL in the night,
someone needed to be close at hand. So without being asked
or getting permission, he put himself on guard duty that night.
He was not going to let anything bad happen to his brother and
that was that.

Diane Cairns

5. Let Your Child be a Child

When you are dealing with diabetes, your whole life is focused around keeping blood sugar levels within an acceptable range. But it is very important not to let this goal of 'keeping things in an acceptable range' spill over into other aspects of your life.

My mother-in-law was always trying to keep my kids in a status quo state in every aspect of their lives. She would say things like "Don't play outside. It's raining out," or "Come into the shade. You don't want to get hot," or "Button up your coat before you get cold." Her idea of being a good mom was to do everything possible to make sure her children didn't ever feel uncomfortable. I understand where she was coming from, but I don't agree with it. I believe if you are never uncomfortable, you can't enjoy being comfortable.

If you never get cold, you can't enjoy the experience of warming up in front of a fire. If you don't get hot at the beach, you won't know how wonderful it feels to jump into the cool ocean. I like loud music so I can enjoy silence. I like chaos so I can enjoy calmness. I can appreciate the fact that not everyone defines 'enjoyment' the way I do. But when it comes to children, 'enjoyment' in all its various forms is embraced and experienced effortlessly. Don't let diabetes interfere with those moments. My tip to you is – **let your child be a child**.

Childhood is a time of wonder. Everything is new and waiting to be experienced. All of a child's senses are alert and responding to an enormous amount of stimuli every day. Interpretations and responses to these stimuli are the foundation of a person's life. Let your child be hot and cold, excited and worried, afraid and exhilarated. Let your child experience life in the unfettered way children do. Encourage your child to experience the ups and downs, the trials and triumphs that come along with living.

When your child is young and you are responsible for his care, this philosophy of living life to the fullest may seem a bit unbalanced. While your child is out there living, you are left with picking up the pieces.

We all know that every time your diabetic child does something out of the ordinary it will affect his BSL and will make life more complicated. You will have to increase something, decrease something else, and monitor something more often. But it is worth it.

I know that it would be easier if he played in a sandbox rather than climbed a tree. I know it would be easier if he stayed home and played with his Lego rather than having gone to a birthday party at the waterslide park. I know it would be easier if he stayed on the shore rather than swam towards the horizon. This list goes on and on. The easier options may be the 'best' with respect to managing diabetes, but the easiest option is not always the best with respect to overall growth.

As the parent, you are setting an example. It is your job to show him that life is meant to be lived. By embracing diabetes and all of the challenges it brings, you are showing your child that life and everything that entails is meant to be enjoyed. After all, what is the alternative? No good can come from being angry or frustrated that diabetes makes everything more difficult. You really don't want your child to grow up with the attitude that he should stay in the sandbox rather than climb to the top of the tree.

Just Sharing –

It very seldom snows where we live. So when it does, it is a big deal. I remember waking my three boys up at 2 o'clock in the morning, so they could go out and play in the snow. I didn't want to take the chance of the snow melting or turning to rain before morning and I didn't want them to miss the opportunity to experience the snow. I knew that hour of activity was going to affect energy consumption, insulin peaks, food absorption, and the following day's sleep cycle. I knew it was going to take many hours of work the following day to get back to 'normal'. Was it worth it? Absolutely!

PS – To this day, my children get excited when it snows, and they are now passing that joy on to my grandchildren.

Diane Cairns

6. Respect the Relationship

The relationship between a parent and a child with diabetes must be based on love, trust and respect. If there is any sort of power struggle or lack of trust on either side, the consequences are immediate and life threatening. There is no room for discontent. It is absolutely imperative that you work as a team but be aware of the fact that the dynamics of your team will constantly be changing: you will never be equal partners. The balance of responsibility is constantly changing.

When your child is under the age of 5, you are totally responsible for their diabetic care. You need to know what they eat, how they feel, their energy level, their mood, what their BSL is and how much insulin they need. You may not have diabetes personally, but you will be living the life of a diabetic. At this stage, you make all of the decisions and they must trust you.

When your child is a teen, you need to understand everything and know what to do, but you really have no control over what they eat or how much insulin they take. You are a supporter and a bystander on call. When they are in their teens, they make all of the decisions and you must trust them.

The balance of responsibilities is most challenging when your child is between 6-12 years old. You will need to be involved in the day-to day-management, but you will also need to be able to stand back and let him make decisions. You have to be very versatile in your role. I have to tell you, this is a little tricky!

When Michael was diagnosed, he was old enough to understand he was responsible for his own health, but too young to do it all on his own. During the time when we were both responsible for his life, we made a solemn vow to each other that we would work together. I promised I would never judge him. He promised he would never lie to me.

Just Sharing –

When Michael was 8 years old, he got fed up with having diabetes and said he wanted to eat 10 chocolate bars. I didn't judge or question his feelings. He didn't sneak out and eat them in hiding. We would work through this together. We decided he should probably take extra insulin before doing it and test his blood frequently for a while after. He enjoyed the first few bars, but he really didn't feel very well while he was eating the rest and decided it wasn't worth it.

When he was 13 years old, he questioned whether he really had diabetes or not. Maybe he was misdiagnosed. He wondered if his pancreas was not making insulin because he was injecting insulin and his body didn't need any more. What would happen if he stopped taking insulin? Would his pancreas kick in and start producing? Knowing Michael as I do, I knew he was going to experiment. Should I have just pulled the old 'I am your mother and I know what is best for you!' line, or were we really equal partners in this journey? The answer was we were partners; so we made a deal. On the day he took no insulin, he would stay close to home, keep me posted on how he was feeling and let me test his blood whenever I wanted to. By the end of the day he had a headache, he was grumpy and he was generally not feeling well. We did some tricky insulin adjustments and injections and made it through to live another day.

He is now an adult and living on his own. I sometimes have nightmares of him going low and of him not being able to help himself. I wake up wondering if I am having 'mommy telepathy' and he needs my help or if I am just being a silly old woman worrying about things I can't control.

I told Michael about these nightmares. As a team we came up with a solution! He told me that anytime I am worried about him I should just phone him – day or night. The worst that can happen is I will wake him up. Big deal – he can just go back to sleep. His promise was that he will always answer the call and not be annoyed – no matter what he is doing!

Regardless of who has what responsibilities, when you both respect the relationship, teamwork thrives and everyone benefits.

Diane Cairns

7. Be Careful of the Truth

We all know that we should 'tell the truth'. But this is not always easy to do, especially when speaking to a young child about diabetes. It is hard to tell a 3-year-old, 6-year-old or 10 year old that "If you don't eat something right now, you will die". But if it is the truth and a reality that your child must understand in order to live, you need to say the words.

The reason I am dedicating a whole chapter to this tip is because the truth with respect to diabetes is not black and white. Diabetes research is happening every day in places all over the world. Diabetes knowledge and therefore diabetes facts are constantly changing. What was 'true' ten years ago is not necessarily true today, and what is true right now may not be true tomorrow. So simply telling the truth is anything but simple.

A person with diabetes needs to know what he is dealing with, no matter what his age. A good rule of thumb when explaining diabetes to your child is to let him ask questions. This way you know what level of understanding he is at and what he is capable of comprehending. Be honest. Every answer must be truthful. Don't ever sugar-coat the facts or bend the truth. If your child finds out the 'real' truth from someone else, your credibility is shot. Avoid this at all costs. Once trust is gone, it is very hard if not impossible to get it back.

I have learned that it is possible to tell the truth if you don't speak to anything past the next 24 hours. "Right now you need to eat something." "Yes you will need to do an injection tomorrow as well." But if you make statements about next week or next year, your chances of inadvertently telling a lie are huge. So if your statement has any reference to the future don't say it unless you precede it with "Based on the knowledge we have today..." and end it with "But that may change next week." For example, don't say things like "You will need to take insulin for the rest of your life." Instead say "Based on the knowledge we have today, you will need to take insulin for the rest of your life. But that may change next year."

The consequences of not using those qualifiers or of basing a future on facts of today can be devastating.

Just Sharing –

My cousin, who is a few years younger than I am, was diagnosed with diabetes when she was eight. She went blind in one eye in her 20s and died from kidney failure in her 30s. Her

son was 2 years old when she died. Michael was in his teens
when she passed away.

After the funeral, Michael and I had an open and honest talk
about diabetic complications and the life expectancy of people
with diabetes. A 'fact' believed at that time, and based on
statistics at that time, was that 'the younger you were when you
were diagnosed, the shorter your life expectancy'. My cousin
was diagnosed at 8 and died in her 30s. Michael was
diagnosed at 6. We can all do the math.

Having a conversation about the long-term health
complications resulting from having diabetes was not a
mistake. The mistake was that in my effort to be honest, I let a
'fact', based on the knowledge of the day, be interpreted as the
truth. Hopes and dreams are what make people get up in the
morning. By believing he was not going to live past 30,
Michael had no reason to dream, to plan, to strive. That
misinterpreted fact, inaccurate truth, lie... whatever you want to
call it, played a pivotal and extremely negative role in
Michael's life. So be careful when you are dealing with the
truth and never let a fact believed today have any role in
defining your future.

8. Forgive

Being able to forgive is an absolutely essential skill to have if you want to be happy. As with any skill, it takes practice to become good at it. Living with diabetes provides you with a lot of opportunities to practice this skill! These 'opportunities' (read as: 'times when something harmed your child and you got so angry you could have spit nails') arise when a decision related to diabetes was based on one of three things: changing rules, following rules to the wrong conclusion or just plain old not knowing something.

8a. Who Makes these Rules?

When Michael was first diagnosed, I was told that if his BSL was high, we should get him to run around and burn up some energy. That sounded logical and Michael was usually willing, so that is what we did. Many years later I was told the exact opposite: that if he exerts himself while his BSL is high he would be causing damage to the small arteries in his eyes, heart and kidneys. The new rule was that when his BSL is high we should give him insulin and get him to sit quietly until his BSL comes down. Hmmm? Had we been causing harm and irreparable damage because we followed the advice of a trusted doctor?

We all know that a doctor would never harm or mislead a patient intentionally. But when your child is harmed, either intentionally or unintentionally, a natural response from you as a parent is anger. What can you do with this anger? Can you be angry at someone who gave the best advice he could, based on the knowledge he had at the time? Can you be angry at someone for changing the rules when the new rules are based on new information? What you can do is use this anger as an opportunity to practice the art of forgiving.

8b. Oops!

Another time you can practice forgiveness is when the rules don't change, but your processing of them does. There has been more than a few times in my life when I thought I was following the rules, but didn't quite connect all the dots and unintentionally ended up making poor diabetic decisions.

I should warn you that the hardest person to forgive is yourself. When you make a mistake and the consequences of your actions could have cost your child his life, it is not easy to just forgive. But you must remember that as a parent, you do a million things every day that are right and probably a few that are brilliant. Give yourself a break. You are only human. Learn from your mistake and move on.

Just Sharing –

One morning when Michael was 8, he was sitting at the breakfast table. He was so cute. He had his PJs on and his hair was messy. He had a serene look on his face as he waited patiently for his breakfast to be served. When I put his pancakes in front of him, I smiled at him and then realized 'He is not quietly being patient. He is going low. I am about to lose him!' My thought process was that he needed food. So I quickly stuffed pancake into his mouth. Now looking back... stuffing food into the mouth of a person who is just about to have a seizure is not the smartest thing to do. I called 911 for help dealing with that one. I had to hand over the reins for a moment and collect myself. I wanted to be angry at myself for reacting incorrectly. I wanted to berate myself for letting my guard down. For a moment I thought of him as an angelic, beautiful child getting ready for a new day and forgot that he was also a diabetic child waiting for breakfast.

8c. New Knowledge

The final type of situation that offers you the opportunity to practice forgiveness is when you find out something you were simply unaware of. For years we had been struggling to get some sort of stability in Michael's BSL. Through research and discussion, we concluded (incorrectly) that Michael was a 'brittle' diabetic. That just happened to be our lot in life and we had to learn to live with it. Then one day we found out that we had been using the wrong type of insulin. Wow – I really wanted to be angry! Why didn't I know about this? Why didn't someone tell me? What kind of society do I live in that doesn't make sure I know things!!

Note: Diabetes has been part of my life for over 25 years and as such, I thought I was an expert in forgiving. But I must confess that when I was writing that last paragraph and thinking about all those challenging times that could have been avoided, I felt a little anger. I guess I need more practice at letting go. But I am sure life will provide me with more opportunities where I can practice the wonderful act of forgiveness.

Just Sharing –

When Michael was a teenager we had too many lows and too many seizures. He had a seizure just before getting into a small row boat to go out fishing in the middle of a lake. He had one just before going on a snowmobile trip. He had one in bed in the middle of the night. He had one in the shower. We had so many that we were getting very efficient at dealing with them. We seldom called 911. We rarely even bothered to tell the doctor.

I am not sure if we should be proud of the fact that we 'handled' them so well or embarrassed by the fact that his diabetes was so out of control. I found those years exhausting. Michael would be back up and enjoying life within hours (even minutes) after a seizure. It was taking me longer and longer to recover from them.

At one point, Michael had three seizures in 3 months. He had the second one while he was in bed. We later determined that he was face down, and during the convulsing his head acted like a big, heavy, bowling ball that just kept smashing into his lower jaw. When I found him there was blood everywhere. Mouth injuries tend to bleed a lot. Once we got the bleeding under control, we realized that his teeth were still in his jaw, but that the front part of his jaw was broken. We went to the Emergency Room at the hospital and got things reattached. Now we were dealing with the challenges of a diabetic who couldn't eat easily. Did it ever stop? We were all getting very worn out. My ability to cope was reaching a dangerously low level.

Diane Cairns

When the third one happened, I called the ambulance. I didn't have the energy to deal with it. In the emergency room, a diabetic specialist was called in and tests were done. The specialist came to us and said "Of course he is having seizures. The type and amount of insulin is all wrong. Try this instead". That was over 10 years ago, and we haven't had a seizure since!

It was as simple as that!! Hmmm? Had all of his seizures been preventable? I was so angry! Should I be angry at Michael? Myself? Doctors? The system? My anger was making me sick. I needed to get rid of it. I decided to look at it as an opportunity to practice forgiveness. I forgave all of the people who played a role in that 5-year fiasco. I learned from the mistake. Then I moved on to being grateful. I was grateful to that doctor for being there in the Emergency Room and sharing the knowledge he had at that time. I am still grateful for the fact that Michael continues to be seizure-free.

8d. The Outside World

So far, all of my examples of forgiveness have been with respect to the medical and physical side of dealing with diabetes. But there is a whole other realm out there that provides even more opportunities to practice forgiveness.

That other realm is the outside world and how it relates to your child. For example, there was a coach who said he would prefer not to have Michael on his team because he didn't want to be responsible for him if he had a diabetic incident. A mother of one of Michael's friends said she didn't mind having Michael attend her son's birthday party, but she wasn't comfortable with him staying for the sleepover part. When Michael was a teen he was with his girlfriend when he started to have a seizure. She started screaming and freaking out. It took more energy to calm her down than it did to get Michael back to normal.

I am not judging any of those people. I can certainly understand how a person may not feel comfortable being responsible for your child's life, and believe me I know how scary witnessing a seizure can be. In each of these situations I did not respond with anger. Instead, I chose to teach, forgive and then move on.

9. Be Thankful

I can't have a chapter on forgiveness without also having a chapter on appreciation. For every scenario that required forgiveness, there were ten that elicited thankfulness and appreciation.

When Michael was 10, he slept over at his friend's house. The mother simply asked me if I could give her a quick rundown on what she should be looking for and what she should do if something happened.

Another time, Michael was starting to go low on the way home from school. His cousin noticed and said "Hop on" and then ran all the way home with Michael on his back.

I remember having a group of 14-year-old boys sitting at my kitchen table wanting me to tell them everything there was to know about diabetes. They wanted to be ready in case something ever happened when they were around.

I can't begin to tell you how grateful I am to my sister and her husband. Right from day one, when my sister was at the hospital with me giving injections to oranges, I knew she was always going to be at my side. Part of my protocol of dealing with a seizure was to call my sister. No matter what time, day or night, she would be at my side within minutes.

A seizure is a scary, traumatic event. Together, my sister and I made sure everyone involved in the experience received the care they needed. Usually, my job was to look after Michael; her job was to look after everyone else (my other children, Michael's friends, etc.). When it was all over, she would look after me.

Just Sharing –

One day when Michael was 15, he did not have to be at school until 10:30, so he slept in and was alone in the house that morning. When he got up at 10:00, he was two hours late having breakfast. As a result, he was low when he got out of bed. He was heading towards the kitchen, but didn't make it. He collapsed in the hallway. While convulsing, his bare foot scraped back and forth across the air vent in the wall. His foot looked like it had been worked over by a cheese grater. We don't know how long his seizure lasted or how long he laid their unconscious, but when he came to, he managed to phone me at work.

When I got the call, I knew something was wrong. He was coherent, but sounded confused. He said there was blood everywhere but he didn't know why. I knew my sister was home, next door in our duplex, so I immediately called her.

When she got there Michael was still sitting at the kitchen table with the phone in his hand. His foot was badly cut and was actually squirting blood. She gave him juice and fed him while she bandaged up his foot. She said my house looked like a war zone. There was blood on both sides of the hall from floor-level half way up the wall. Then there was a trail of blood into the kitchen, pooling around the base of the fridge and the area by the phone.

When I got home 30 minutes later, the walls and the floors were clean. Michael was looking very normal, putting on his shoes and getting ready for school. My sister, however, was looking a bit dazed. On this occasion she looked after Michael. I looked after her.

10. Take Care of Yourself

If you have ever been on an airplane you will know the safety routine they demonstrate. They always tell you to put your life jacket or oxygen mask on yourself first – then take care of your children. The reason they have to take the time to repeat this message each and every time people enter an airplane is because taking care of yourself first is not a natural thing for a parent to do. A parent's natural response is always to take care of their child first. Once a parent is sure their child is okay, then they take the time to care for themselves. I want to tell you that the airlines have it right! You can't take care of your child if you are not there (either physically or mentally). You need to be healthy and strong in order to have the energy and ability to take care of your child.

The demands of caring for a diabetic child are never over. There is never a time when you can wash your hands and say "There we go. My job here is done." You are always listening, preparing, planning, learning, or on standby waiting. My tip to you is 'intentionally schedule time to take care of yourself'. Don't just wait for a break in the action. It doesn't happen often enough! You must pass the reins over to someone you trust and then walk away. It can be as simple as locking the bathroom door and having a bubble bath (a long, hot bubble bath with a good book and a cup of tea!) or as major as leaving town (to a 5-star Resort and Spa!) for a few days.

Just Sharing –

My mother died, I had a miscarriage, my son was diagnosed with diabetes and my husband left me all within a five-year period. I had debilitating headaches that lasted for 10 days at a time. I was exhausted. I couldn't walk from one end of the house to the other without sitting down. I cried every day.

One morning, I asked my sister if she could look after my youngest son while I went grocery shopping. My two other children and her three children were all at school and she was happy to look after my little guy. When I left the house my intention really was to go shopping, but when the turnoff to the store came, I didn't turn in. I just kept on going. The entrance to the freeway seemed to be calling to me. I got on the freeway and just kept going. I had driven for about two hours when the tears in my eyes were making it difficult to see the road in front of me, so I pulled off and found myself in a community where my cousin lived. When I got to her house, she took one look at me, called her husband at work and asked him to come home to look after their kids. She told him she had to take care of me and wasn't sure what that entailed or how long it would take.

We ended up going out of town for a couple of days. What a wonderfully understanding and generous family I have! I am very fortunate to be surrounded by such loving, non-judging human beings.

After my third 'runaway', my sister (and her husband), my cousin (and her husband) and my husband very politely asked if I could give them a bit more notice when I was going to run away. Since that seemed like a reasonable request, I did start to plan my runaways (now called getaways) in advance. I did two out-of-town, overnight getaways every year. The benefits of those self-care breaks were invaluable. I got to be a person (not a mother, not a caregiver, just a person). I was a woman who was enjoying being alive and healthy. I could take the time to see the world without plans. I could turn left or right. I could sit or stand. I could even take a nap, just because I wanted to! Now, twenty-five years later, my cousin and I still go on two mini vacations every year.

Scheduled self-care breaks (short or long) prevent you from getting exhausted and reaching the point where you just can't do it anymore. When you know there is a light at the end of the tunnel, it is easier to keep going.

Diane Cairns

No matter how often or what method you use, self-care should be a priority for everyone. Just like the airlines say, you need to put your oxygen mask on first. You never know when life is going to hand you something that will require all of your energy. You will need to be ready.

Conclusion

Have you ever had the wonderful experience of having someone effortlessly help you? I have, and I can tell you it is a beautiful thing!

I was 20 years old and had been trying to get a crying baby to stop crying. Nothing I did worked: rocking, rubbing, patting, feeding, ignoring, cuddling... The little darlin' was just getting more and more wound up. My next move was to just join in and start crying myself. That was when my grandmother entered the picture. (To this day I am not sure why she arrived at that moment, but I am glad she did!) She took the baby, stepped out onto the sundeck, held the baby against her chest and started singing. Whether it was the change in temperature, scenery, or voice, I don't know, but the baby just stopped crying, looked around, put her head down, let out a huge sigh and settled into a very relaxed and well-deserved sleep. What a blissful moment.

It happened more recently as well. I was trying to get my computer to do something I knew it could do, but no matter what I tried, it just wouldn't do it!! My daughter-in-law came over and simply said `Have you tried this?`. Ta da! It worked. Oh what a feeling!

Diane Cairns

That is the feeling I want this book to give you.

From one parent of a child with diabetes to another, I hope that by sharing my top ten survival tips with you, I am relieving some of your frustration, helping you to find a moment of peace and increasing your ability to succeed.

PS - If you would like to send me a message, I can be reached at bdcauthor@gmail.com.

Take care,

Diane

www.ingramcontent.com/pod-product-compliance
Lightning Source LLC
Chambersburg PA
CBHW021917170526
45157CB00005B/2095